All About Wind

by

BethBirdBooks

THIS BOOK BELONGS TO

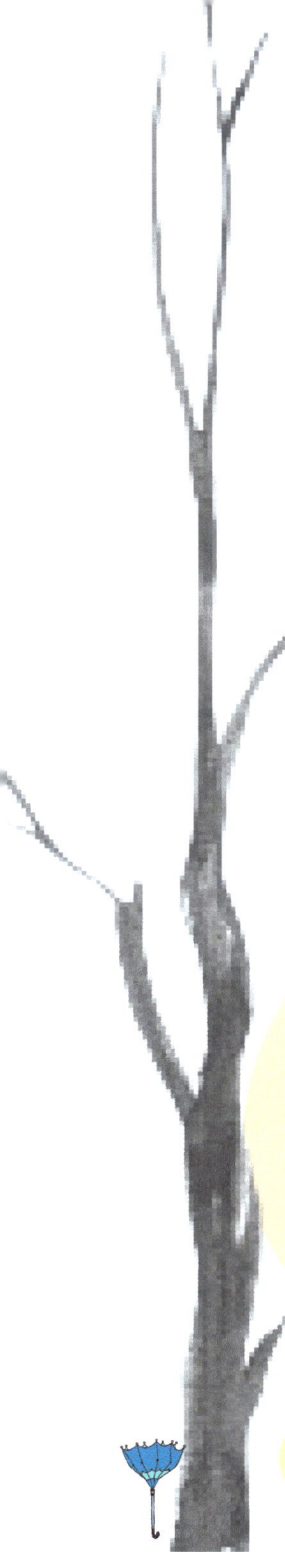

_____

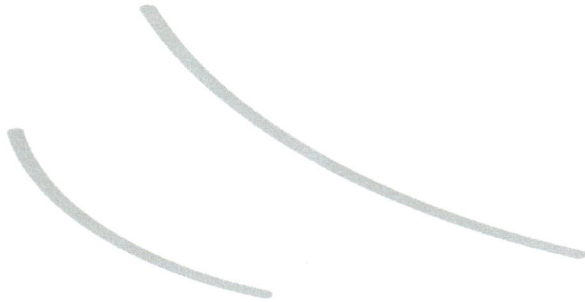

Wind is moving air. We cannot see the wind,
but we can see what it does.

Sometimes you can see the wind move clouds across the sky.

Tree tops sway in the wind,
and leaves blow in the air.

Wind chases leaves through the yard.
Wind can make raking leaves harder.
Using a leaf blower may be easier.

Wind causes flags to flap in the breeze.

A pinwheel spins when you blow on it,
or move it quickly through the air.

Birds spread their wings and soar on wind currents.

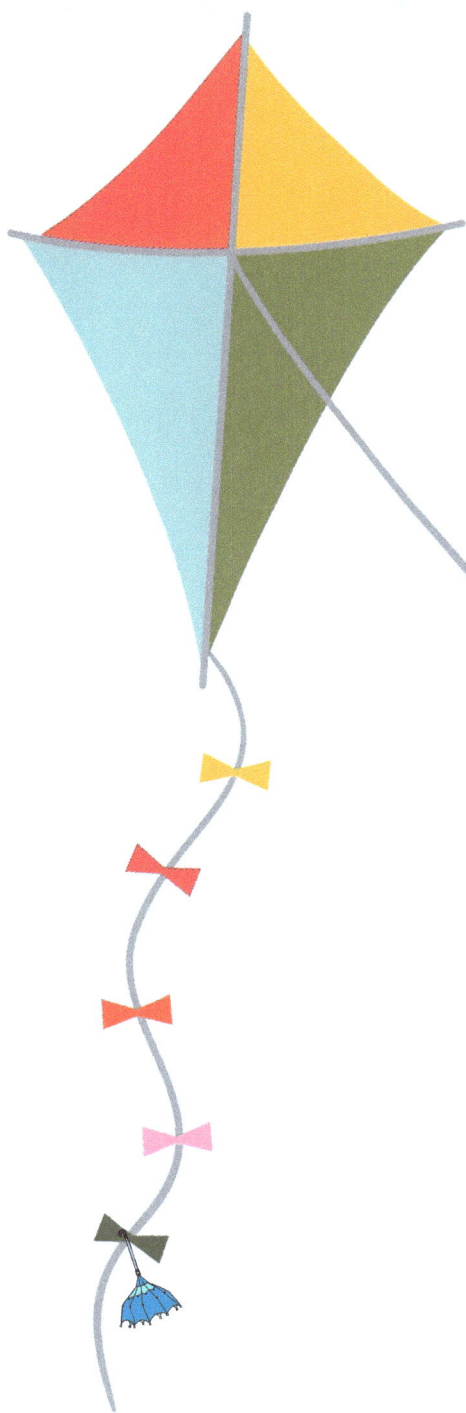

The wind lifts your kite in the sky.
You might make it go higher
by running as you pull it along.

The wind spreads seeds for new plants to grow.

Many musical instruments use moving air to make sounds.

When you cough or sneeze, it moves the air. A sneeze can travel as fast as 100 miles per hour.

To keep from spreading germs, turn away from other people and cough or sneeze into your elbow.

A farmer checks the weathervane
for wind direction.

A windsock shows which
way the wind is blowing.

The sun heats the Earth
unevenly, causing areas
of high and low pressure.

When hot air rises it creates low air pressure. Cooler, heavier air moves in to take its place causing high air pressure. This moving air is wind.

Wind can be helpful. For centuries people all over the world have used the power of the wind. The first ships sailed the ocean by wind power.

We still use wind to power sail boats.

Wind power can turn a heavy
millstone to grind grain into flour.

You can bake bread and yummy treats with flour.

A farmer might erect a windmill to pump water from underground for his family and livestock.

Today, wind farms on land and offshore
harness wind energy for electricity.

A fan creates wind to help you feel cooler.

A breeze can help you feel cooler on a sweltering day.

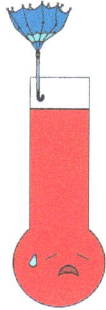

On cold days the wind can make you feel even colder.
This is called wind chill.

When the wind is blowing hard, it can turn your umbrella inside out!

Sometimes wind is scary. Storms can cause
strong winds, like tornados and hurricanes.

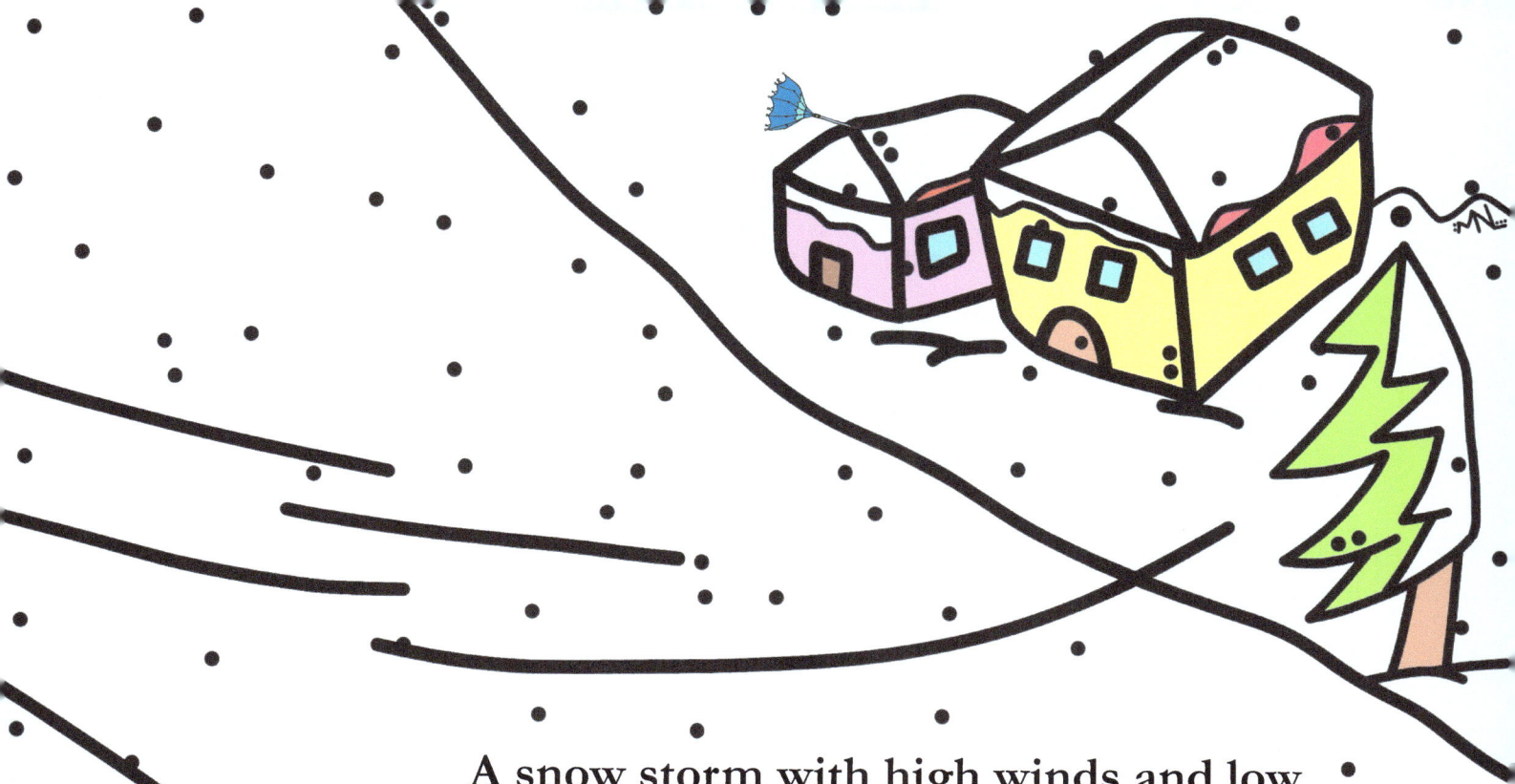

A snow storm with high winds and low visibility is called a blizzard.

Wind can bring enormous clouds of sand or dust.
Dust storms are more likely when there hasn't been enough rain.

Wind can spread dangerous wild fires.

It is not safe to be outside when it is too hot or cold,
or in a tornado, hurricane, blizzard, dust storm, or fire.

The weather forecaster lets us know
when extreme weather is expected,
so we can stay safe.

How are these three things alike? How are they different?

They all have turning blades.
A fan makes wind, but blades of the pinwheel and
windmill are moved by wind.

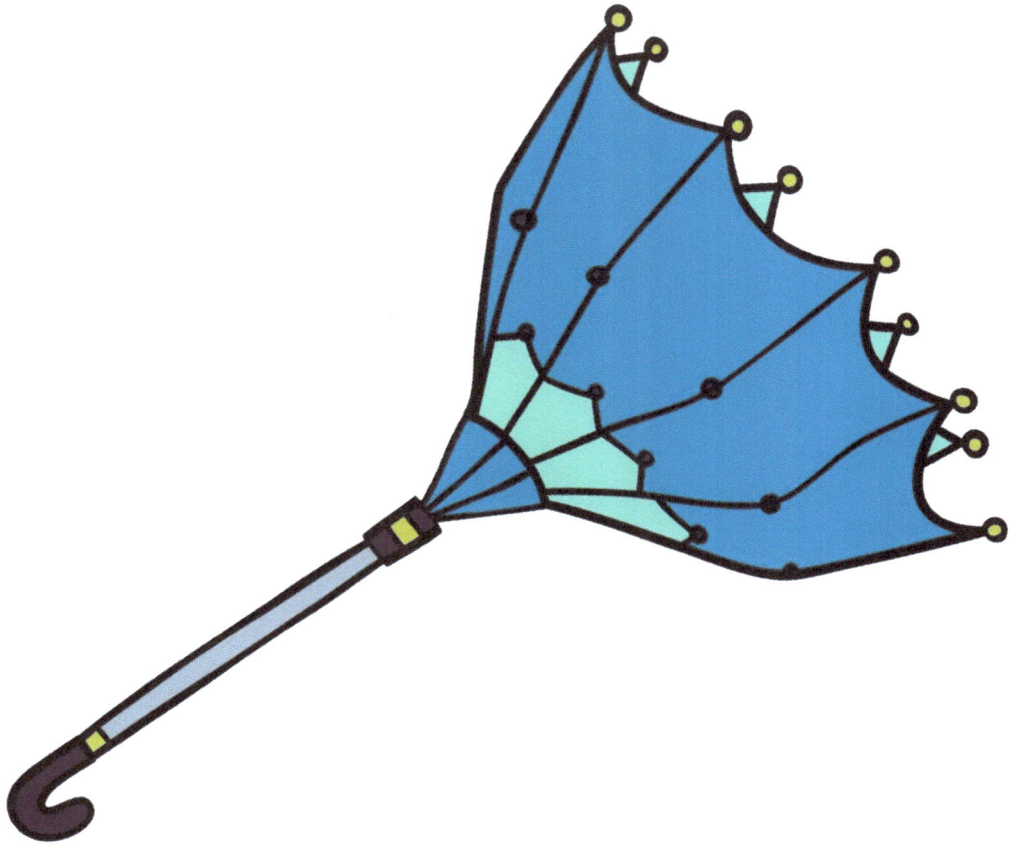

The End

www.ingramcontent.com/pod-product-compliance
Lightning Source LLC
Chambersburg PA
CBHW052048190326
41521CB00002BA/153

9781950603138